BEYOND THE REACH

CRADLE MOUNTAIN–LAKE ST CLAIR NATIONAL PARK

BEYOND THE REACH

CRADLE MOUNTAIN–LAKE ST CLAIR NATIONAL PARK

CHRIS BELL

Published and distributed by Laurel Press
PO Box 132, Sandy Bay, Tasmania 7005 Australia

All rights reserved. Except for the purposes of review, no part of this book may be reproduced in any form or by any means without prior permission in writing from the publisher.

Photographs and text copyright © 1990 by Chris Bell

Designed and wholly produced in Australia
by Rodney M. Poole Pty Ltd

ISBN 0 646 01323 8

Acknowledgements

Many people have assisted in the production of this book. In particular I would like to thank the following: Rex Direen, Stuart Graham, Mark Errey, John Hughes and David Bell who assisted in the logistics of being able to spend long periods in the mountains; Martin Hawes, Sam Rando, David Ziegeler, Rob Blakers, James Warden, Geoff Lea, Chris Sharples, Peter Dombrovskis and Liz Coombe for their helpful criticism of the text, and to Alan Haig for his editorial assistance.

I would also like to thank Greg Buckman and John Langford for assisting in getting the project off the ground; Ian Skinner for the map, Dave Watts for the two wildlife photographs, Bob Brown for his overall and ongoing assistance and to Linhof Prazisions-Kamera-Werke GmbH in West Germany.

Finally, a special thanks to my constant companion in and out of the bush, Jenny Burnett.

Contents

A NATURAL HISTORY

THE PARK AND ITS BEGINNINGS 8

GEOLOGY AND THE LANDSCAPE 14

FLORA AND FAUNA 20

WILDERNESS 28

WILDERNESS: A PERSONAL ACCOUNT 35

THE PORTFOLIOS 39

EPILOGUE: A WESTERN TASMANIA NATIONAL PARK 88

About this book

This book is a memento of Cradle Mountain–Lake St Clair National Park, the outstanding temperate landscape which covers much of Tasmania's Central Highlands. The book is about Nature, and in particular about its wildest face, wilderness. It is not intended to be a detailed study of the park, but rather a celebration of its hidden beauty.

This is, ultimately, a photographer's book, so naturally its core lies in the photographs. Chapters on geology, flora and fauna, and so on, are included solely for background reference; much of the text comes closer to being a personal statement about wilderness, and about Tasmania's wilderness in particular. Further, I have avoided addressing the management dilemmas which loom ahead for this park and wilderness in general. While these issues are crucial they are also, I believe, beyond the scope of this book.

A word on the photographs. I have deliberately not disclosed the specific locations of some of the photos; while this may seem unorthodox to many, I wish to avoid the trap of enticing people to secluded or fragile areas, or of raising their expectations. I maintain that the places which will inevitably mean most to us are the secret places we discover for ourselves. Cradle Mountain–Lake St Clair National Park is full of them.

As the world accelerates towards a more unsustainable future, humans are increasingly losing touch with the natural world, and the wild face of planet Earth is disappearing accordingly. Country that was wild in recent memory is now ploughed, settled or stocked; there is not much time to safeguard Earth's remaining treasures. Several organisations such as the Australian Conservation Foundation and the Wilderness Society are working to secure our wilderness; with your participation we stand a chance.

A NATURAL HISTORY

THE PARK AND ITS BEGINNINGS

CRADLE MOUNTAIN AND DOVE LAKE, SUNRISE

The Park and its Beginnings

Cradle Mountain–Lake St Clair National Park is essentially a wilderness area. It is largely remote and undisturbed, one of the last such temperate landscapes remaining in Australia. The boundaries encompass deep, forested valleys, high open moors and ancient craggy peaks. Some of Australia's finest mountain country is to be found here. The soaring cirque walls of fluted dolerite, thousands of lakes, tarns and pools, the wild open plains and U-shaped valleys, all reflect the park's glacial past. Some of the most intact temperate forests remaining in Australia, forests of ancient lineage, are protected within these boundaries. Many of their species are restricted to Tasmania.

Four of Tasmania's largest rivers — the Derwent, Franklin, Mersey and Forth — begin their journeys here, born of icy tarns and lakes that record some of the highest rainfalls in Australia.

The high country of Tasmania, and in particular the Central Highlands, is characterised by a phenomenon almost non-existent on the Australian mainland — four defined seasons. In autumn, under calm, soft-lit skies, the leaves of Australia's only winter-deciduous tree, the endemic Deciduous Beech, turn golden orange before scattering with the winds and snow of winter.

This area is home to many unique animals and birds and in the past was home also to the dark skinned original inhabitants of Tasmania. The Aborigines first occupied Tasmania around 30 000 years ago, before the onset of the last ice age. Over the next few thousand years the climate began to cool, transforming the land into a different landscape from today's: large valley glaciers poured from ice-caps which covered the Central Plateau and parts of the Central Highlands.

Aborigines lived on the fringes of the ice for thousands of years. During this time the open plains and foothills constituted a mosaic of herb-fields and alpine grasses, arctic in character. It is most likely that the Aborigines restricted themselves to the milder, more sheltered valleys, which were free from the ravages of ice and snow and supported ample game.

When the Earth's climate began to warm, about 12 000 years ago, the ice-caps shrank, taking the valley glaciers with them. The last great ice phase was ending. The rainforests, previously confined to the lower valleys with their milder conditions, began to follow the retreating ice and to colonise the deglaciated land.

As the forests advanced into the ice-free areas, so too did the Aborigines, and they brought with them the most ancient of human tools — fire. They used fire to encourage game, and over the next few thousand years this altered the distribution of Tasmania's vegetation: rainforests retreated and the more successful modern plants took their place. The present mix of eucalypt and button-grass communities is largely the result of this fire regime.

Little evidence remains of the Aboriginal presence within the park. But the few known sites, scattered throughout the park, confirm that they visited or occupied many areas during the next 10 000 years.

By the time the first Europeans started to penetrate this wild interior, the Aborigines all over Tasmania were being hunted by the new settlers and soldiers; their bands began to disintegrate.

In 1832 W. S. Sharland became the first European to gaze upon the lake the Aborigines called Leeawulena – the "Sleeping Water". Sharland viewed the lake from a distant hill but did not reach its shores and it was the surveyor George Frankland who finally did so in 1835, and who named it St Clair.

By this time the original inhabitants, who had seen the Tasmanian landscape in transition, were almost gone. While their descendants survive, the original race has vanished — victims of encroachment and misunderstanding by invaders who would take years to come to love this land. Among those invaders were a few fur-trappers and prospectors who tried to extract a living from this saturated wilderness; a few of their relics, not yet overgrown, are visible still.

Among the first Europeans to appreciate the place —

CLEARING STORM, LAKE ST CLAIR

and perhaps the person whose name is most closely associated with the park's establishment — was an Austrian, Gustav Weindorfer. Weindorfer was born in 1874 and migrated to Australia in 1900. He was a keen botanist and made many excursions collecting plants in Victoria, before he moved to Tasmania. While climbing Mt Roland, on the northwest coast, he was taken by a distant mountain that would later shape his life. It was Cradle Mountain, which rears abruptly from wild moors and cool forests, a dominant feature of the northwest. Weindorfer visited Cradle Mountain for the first time in 1909, and camped on the shores of Lake Dove. It was the start of his life-long attachment.

Weindorfer returned in 1910 and reached the summit. Gazing out at the romantic vista of distant wild peaks, with outstretched arms Weindorfer proclaimed his now-famous words: "This must be a national park for the people for all time. It is magnificent, and people should know about it and enjoy it."

In 1912 he built his forest home — "Waldheim" — and in 1917 he finally settled there, remaining until his death in 1932. In those years Weindorfer was the genial host of Waldheim, and it became a welcome retreat for many of the area's early travellers.

In 1922 a scenic reserve was declared, covering 64 000 hectares. After 1940, when modern bushwalking became popular, significant numbers of people began venturing here for reasons very different from those of the early pioneers: for the love of wild, remote country.

A number of later additions enlarged the park to 132 000 hectares, and in 1982 the park was inscribed on the World Heritage List, along with the Southwest National Park and the Franklin–Gordon Wild Rivers National Park. In 1990 the boundaries were again enlarged, and simultaneously rationalised. It now covers 161 000 hectares.

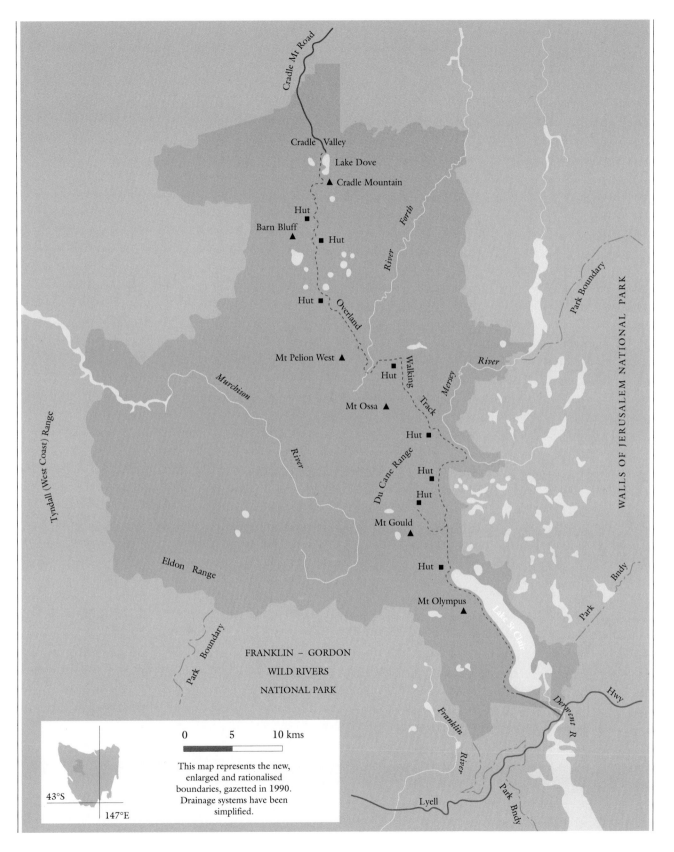

GEOLOGY AND THE LANDSCAPE

PERMIAN FOSSILS IN MARINE SILTSTONE, BARN BLUFF

Geology and the Landscape

As you gaze out at the crumpled horizon of broken-edged peaks, you cannot help sensing that these romantic, weathered mountains are old. The jumbled protrusions of dolerite look as though they have been here forever. Like all mountains, of course, they have not; the pink-tinged boulderfields and the defiant columns of leaning stone are but remnants of a much more extensive crust worn away over time by the persistence of the elements.

What you see here is superficial. There is far more to these ancient mountains than is visible. They protrude from, or overlie other and even more ancient rock, rock so old that it contains no "pictorial" evidence of life, for it dates from the Precambrian era when life was just beginning. Much of western Tasmania consists of such rock. When you traverse the quartzite shoulder of Hansons Peak beneath Cradle Mountain, you are walking on rock one-fifth the age of the Earth.

Overlying these Precambrian quartzites and schists, and forming a base for most of the major mountains, is an association of mudstones, sandstones and conglomerates from the Triassic and Permian periods (between about 290 and 195 million years ago). It is these rocks that contain fossils, those fascinating records of prehistory: imprints of creatures living when these rocks were sludge, and which survive as images in stone.

These mountains are distinctive. There is nothing remotely like them anywhere on the Australian mainland; the high alps of Victoria and New South Wales are gentle and rolling, and lack the spectacular landforms of these ranges. And New Zealand, relatively close by, may show many similarities in its vegetation, but that is all. Its mountains are totally unlike these; they are of greywacke, granite, chert and gneiss, and are still high enough to sustain their glacial aprons.

The igneous dolerite that characterises much of eastern Tasmania, the Central Plateau and the Central Highlands, intruded the earlier foundation rocks as dykes or sills in the Jurassic. This raised mass of igneous rock underwent 180 million years of weathering before the onset of the Quaternary ice.

While the older periods determined the structure of these mountains, it was the most recent epoch, the Pleistocene, which most dramatically shaped them. During this period about one-sixth of Tasmania was cloaked in ice. The mountains visible today were then almost totally submerged under this silent shroud; only the highest summits protruded above the grinding ice which shaved away the excess rock. For 10 000 years these peaks presided over the frozen expanses as the ice worked methodically on the submerged geological surface.

At the height of the Pleistocene huge valley glaciers radiated like spokes from the ice-caps; the longest of them, it is now believed, reached almost to the sea. These particular glaciers — the Mersey, Forth and Leven — slid on their beds of pulverised rock for over 100 kilometres, to terminate in what is now the rich farmland of the northwest coast. Others, like the 40-kilometre-long Narcissus Glacier, chewed away cirque-cliffs from Mt Massif, Geryon and the Acropolis before preparing the foundations for the "Sleeping Water". Subsidiary glaciers, like the Pine Valley Glacier, trenched U-shaped valleys.

The glaciers disappeared between 15 000 and 10 000 years ago, leaving a fascinating array of landforms. Time and ice, then, have crafted these mountains, each with its own distinctive features. Sometimes the summits are narrow ridges of toppled columns — leaning towers of shattered stone, barely negotiable; often they are more gentle, culminating in exposed tiers and shelves that slope and drop into conjoining pools and tarns. Soaring headwalls of fluted dolerite, some of them 300 metres high, and U-shaped valleys dominate the landscape now.

As modern Tasmanians we can only imagine what this land looked like in the Pleistocene, and wonder at the astonishing forces of this glaciation; yet there were people who saw it and lived with it. Aborigines, over hundreds of generations, would have heard the thunder

CIRQUE HEADWALL, WINTER, DU CANE RANGE

of avalanches and the growl of moving ice. They witnessed the advance, retreat and eventual disappearance of these mysterious frozen rivers. Over several thousand years, they experienced an extraordinary chapter of Tasmania's history which no European ever saw, or is ever likely to see. Their impressions are lost to us.

When the ice finally melted away countless sparkling lakes remained, many of them doubtless playing a role in Aboriginal legend. One of these — their "Sleeping Water" — plunges to over 200 metres, and is the deepest lake in Australia. This is Australia's lake country, as well as its mountain country. It is estimated that there are over 4 000 lakes and tarns on the Central Plateau alone.

Complementing the mountains and lakes, the rivers constitute the third dominant feature of the park's landscape. Along some of the ice routes surge today some of Tasmania's longest, wildest and least-known rivers. Some finish as major attractions of our national parks; others, sadly, as victims of electricity generation. Of all the major rivers that rise in the Central Highlands, only one — the Franklin — still flows the way it should, as a breathing strand in a living land.

Cradle Mountain–Lake St Clair National Park is a place to contemplate; a place to observe the forces of Nature that still govern this land. It provides a starting point for us to acquaint ourselves with wildness; it elevates our spirits and, hopefully, stirs our consciences to take zealous care of this priceless treasure.

MT HYPERION, SUNSET

FLORA AND FAUNA

MILLIGANIA LILY, MOUNTAIN ROCKET AND SUNLIT PENCIL PINE

Flora and Fauna

To put the park's special flora and fauna into context, even briefly, it is worthwhile to begin our story in the Cretaceous, for this is the era in which our vegetation and wildlife have their origins.

For about the first half of the dinosaur epoch, conifers and ferns — antecedents of extant Tasmanian species — dominated the Earth's vegetation. Then, about 120 million years ago, the first ancestral flowering plants began to evolve and to spread throughout the supercontinent of Gondwana, of which Australia was part. This critical evolutionary moment represented the change from ancient-style to modern-style vegetation. Together with the subsequent break-up of the ancient landmass, it changed forever the vegetation, wildlife and indeed the whole character of planet Earth.

When the Gondwanan continents split apart and drifted north on their associated crustal plates, they were all carrying with them the progenitors of this newer vegetation — the modern flowering plants, or Angiosperms. The rapidly evolving Angiosperms were so successful at adapting to changing climates that they soon began to dominate the Earth's vegetation. By the mid-Tertiary (40 million years ago), the ancient-style Mesozoic vegetation — the conifers, ferns and cycads of archaic stock — had been largely overtaken by the advance of the still evolving Angiosperms.

Over the next 10 million years Australia's climate became increasingly arid and the "transitional vegetation" — the primitively flowering rainforests of the Tertiary period — also succumbed to the more modern Angiosperms, which were now evolving to cope with the drying continent. By the time Europeans arrived, only fragments of rainforest remained on the hot, dry Australian mainland — less than 1 per cent of the landmass. Exploitation for timber and clearance for farming have now reduced it to 0.25 per cent.

Tasmania, by contrast, still supports large areas of its Gondwanan forests, and perhaps the richest examples are found in Cradle Mountain–Lake St Clair National Park. The forests here contain the descendants of some of the oldest known vegetation. Among this vegetation is the ancient lineage of Nothofagus, the southern beeches. This 65-million-year-old genus is found today in southern South America, New Guinea, New Caledonia, New Zealand, a few scattered localities on the Australian mainland — and Tasmania.

A few of the species common in the Tasmanian rainforest, such as Myrtle (*Nothofagus cunninghamii*) and Sassafras (*Atherosperma moschatum*), also occur in small pockets on the Australian mainland; but many are confined to Tasmania. It is these endemics which give Tasmanian forests their distinctive primeval character. Some of the endemic rainforest species seem more akin to the deciduous forests of Patagonia than to anything in the Australasian region.

Outstanding examples of these primeval forests can be found on the northern and eastern flanks of Cradle Mountain. Here, endemic King Billy Pines (*Athrotaxis selaginoides*) and Pencil Pines (*Athrotaxis cupressoides*) tower over the gnarled and twisty Deciduous Beech (*Nothofagus gunnii*). In autumn, the golden-orange blaze of this endemic tree presents one of the most colourful spectacles found in the park. Understoreys of endemic Celery-top Pine (*Phyllocladus aspleniifolius*) and flowering Leatherwoods (*Eucryphia lucida*) add to the complexity of these forests.

As well as its ancient-style vegetation, the park has a rich variety of the modern flowering types. There are extensive forests of eucalypt, which include the world's tallest flowering plants. Among the tallest of the eucalypts, and common in this park, is the Gum-topped Stringybark (*Eucalyptus delegatensis*). Some of the finest examples of these forests, with associated beech forest, can be found in the Forth Gorge, recently included in the park.

About 50 per cent of Tasmania's alpine vegetation is endemic. A large number of these shrubs and flowers are found in this park. Showy examples include the Tasmanian Waratah (*Telopea truncata*), dotting the slopes with crimson, and the Native Laurel (*Anopterus*

DECIDUOUS BEECH, PENCIL PINES AND KING BILLY PINES, AUTUMN

WHITE GOSHAWK   PHOTO BY DAVE WATTS

TASMANIAN DEVIL   PHOTO BY DAVE WATTS

glandulosus), with its cluster of soft cream flowers. There are lovely displays of Gentian sp. and of Mountain Rockets (*Bellendena montana*), and the beautifully scented flowering heads of the Milligania Lilies (*Milligania densiflora*) flourish in the alpine gullies.

Growing in almost all the varied habitats is perhaps the most distinctive of Tasmanian plants, the Pandani (*Richea pandanifolia*). This wondrous, ancient-looking plant is found from sea-level up to all but the highest alpine zones.

The fauna of Tasmania also shows its Gondwanan origins. The mammals consist largely of marsupials; the only terrestrial placentals are bats, and native rats and mice. Included in this array of mammals are two of the world's most fascinating animals, the monotremes, or egg-laying mammals.

The Tasmanian Echidna (*Tachyglossus aculeatus*) shows evolutionary adaptations to Tasmania's cool climate; unlike the mainland version it has, in addition to its spikes, dense fur. These little animals can be seen almost anywhere in lower levels of the park.

The Platypus (*Ornithorhynchus anatinus*) is one of the most extraordinary furred creatures on Earth, with its unique combination of a duck-like bill, webbed feet, beaver-like tail, poison spur on the hind feet, and eye shutters which enable it to swim blind under water. It is still common in Tasmania, although its habitat on the mainland has often been either destroyed or severely degraded.

While all these animals are distinctive and attractive, it is the two endemic marsupials which — more than any other Australian animals — have created a legend: the Tasmanian Devil and the Thylacine. The Tasmanian Devil (*Sarcophilus harrisii*), a stocky black and white animal which has the unique ability to devour all its prey — skin, bones and fur — is still common in this park and can be readily observed in Cradle Valley.

Somewhere in this park, or elsewhere in a similar, remote habitat, the Thylacine (*Thylacinus cynocephalus*, popularly called Tasmanian Tiger) may be holding on to survival. Several reputed sightings of the animal, last indisputably recorded in 1933, suggest it may still exist. If so, it is the remaining wild areas that will hold the key to its survival.

For its size, Tasmania is also relatively rich in birds. Over 200 species are found here, 11 of them endemic. Most of these inhabit the park.

For the naturalist, then, Cradle Mountain–Lake St Clair is a special place. To observe a goshawk on the heights, or a grebe riding the ripples of a wind-whipped pond, is as uplifting as to see the clouds part to reveal the beckoning mountains. These mountains have so far remained apart and remote; if people can appreciate their spiritual significance, they will always be so.

CUSHION PLANTS AND SCOPARIA, MT OSSA

WILDERNESS

MT PELION WEST AND MT OSSA, SUNSET

Wilderness is a large tract of original landscape which is essentially devoid of human alterations, and sufficiently distant from human activities to impress the feeling of remoteness.

It seems incomprehensible that the wilderness — which once fed us and clothed us, and which still comforts and inspires us — has been so abused. For almost all the 3 million to 5 million years of human evolution we displayed deep attachment to our wilderness; now we see only its potential material wealth.

The providing Earth should be our most sacred and healing resource; yet we have seemed bent on extracting more and more from our finite garden and loving it less and less.

Several years ago the vista from the summit of Mt Ossa, in the middle of this park, seemed boundless. Distant valleys lost themselves in a maze of untouched forest; wild rivers rushed through seldom-visited gorges, beneath remote hills that had *never* been visited. The wilderness seemed more intact and secluded than anywhere on Earth; the mountains rolled on, fold after fold, wall after wall. The hazy light, suffusing valleys awash with birdsong, created a sense of dreamy spaciousness, luring you further into the illusion. It was very different then, and really did seem to go on forever.

But the modern world has now intruded upon these open spaces. From the summit, if you look carefully, can be seen the evidence — however small — of the things this park and its wilderness do not want: the distant scar of the badly located and poorly constructed Hydro-Electric Commission road on the west coast, the disastrous high altitude logging coupe on Clumner Bluff to the east, and the evidence of escaped fires. All are the result of careless planning; with a little foresight they need not have been visible, and probably need not have been there at all.

Towards the west flows the Murchison, one of the wild and frothing rivers that begin in this park. Only a handful of people can say they knew the river, and their number will never grow; for further west squats the concrete wall which distorts and disfigures the river's flow — justified by engineers using the same stale rhetoric that engineers use about rivers the world over. Perhaps as few as a dozen people ever saw this section of the river before the flooding.

What chance then, does our wilderness have? What would the world be like without such places?

When Tasman sailed cautiously down the forbidding west coast in the seventeenth century, all he knew about his discovery was that it was mysterious. "Terra Australis incognita". No European had sailed this way before. Perhaps equally disconcerting was the sight of little curls of smoke which drifted from the shore: people were here.

He could not know that they had been living on this island long before the settling of the modern European peoples in their own distant homelands. They had been living in this isolated Eden for over 30 000 years; when they were later dispossessed, most of it was still an Eden. Some people live with a land and are "remote" from it; others can live with a land and *become* the land.

At a time when Europe was being irrevocably altered, Tasmania was still a paradise. But Europeans would see it as simply another field to reap. When they began colonising Tasman's Van Diemen's Land, they brought with them the common stock of attitudes to the land which already prevailed in Europe, and persist almost everywhere to this day: that the land was to be used for whatever need humans saw fit, and existed for humans' sake alone — though not for dark-skinned humans' sake, but for that of those "cultured, educated people" whose attitudes were to transform the treatment of this land. For the land would now be worked, rather than left idle. (One hundred and fifty years later they would distort this ethic even further and talk of rivers running "wasted" to the sea.) The new immigrants would fear, despise and persecute the original inhabitants who seemed to "do" nothing with the land. Europeans planted exotic trees and tried by every means available to turn what was here into something which was familiar

SNOW SHOWERS, DU CANE RANGE, WINTER

to them — to create a replica of the places from which many had been expelled: the safe, green cow paddocks of England and Ireland.

Few saw value in the sodden wildland; even fewer ever sought to conserve any of it; almost no one found solace in it.

Nonetheless, as late as 1950 a good deal of the island was still much as Tasman saw it when he stepped ashore in 1642. In 1950 vast tracts of lush primeval forest sprawled unhindered, most of it unmapped and much of it still quite unknown. Most of the rivers were still living.

Perhaps 1967 — the year the Gordon River Road and the Scotts Peak Road sliced into the southwest — was the turning point for our wilderness. Since then the intrusions have accelerated.

Remote landscapes are becoming increasingly rare in our world. The wild, raw, original face of our island — of our planet — is being lost as societies become more and more preoccupied with the workings of economies rather than with the workings of our environment. There are few places today where we can immerse ourselves in large tracts of unknown country, but we still have that opportunity in Tasmania.

Cradle Mountain–Lake St Clair National Park, together with the wild and remote country surrounding it, is one of our last wild areas, and it represents one of our last chances to adopt a new style of care for our enchanting island, a concern based upon more than mere sustainability. If the human race is to endure we will need to go beyond concepts of mathematical sustainability: we will need to love our wilderness again.

But such bonds are established slowly. People have to learn to love this wilderness. Conditioning, education and the distractions of modern life have made big inroads on our minds, increasingly separating us from the natural world. In our out-of-touch world, the catchcry of "jobs" almost drowns out the new-found concern for our wild country, and convinces many to argue for wilderness on the basis that it will mean revenue for Tasmania. To defend wilderness on these grounds is to argue on the wrong level. To assign monetary value to wilderness is to debase it. We cannot measure solace or freedom in cash terms.

Australia lags behind. We still do not have a Wilderness Act, whereas the USA has had one for a quarter of a century. There has been no authoritative wilderness inventory on a federal level. Wilderness is still seen by many as an obstruction — at best as a commodity — rather than as something more profound. Wilderness preservation is about love of the land. How we rate the effect it has on us depends on whether we see with our hearts as well as our eyes.

Tasmania's most precious resource lies not in the mineral and forest wealth which we exploit, but in the raw, untainted, original face of the island. Wilderness endures; wealth is ephemeral, and always will be.

SUNSET ON PENCIL PINE, WINDERMERE PLAINS

WILDERNESS: A PERSONAL ACCOUNT

Wilderness: A personal account

> *"The demagogues...who have already caused the death of several civilisations, harass men so that they will not reflect; manage to keep them herded together in crowds so that they cannot reconstruct their individuality in the one place where it can be reconstructed, which is in solitude."*
>
> *Ortega y Gasset*

From the top of the mountain I can see clear into forever. In the creamy, diffused light there is no horizon. Sky merges into land, the distant mountains float.

Scattered across the pavements of fractured rock, out-of-season gentians spring from sheltered cracks, withered stragglers hanging on to the last, as gentians do. Far below, another world away, the valleys tinkle with the chirruping calls of the Crescent Honeyeater. With its black breast-crescent, yellow wing-patch and spirited calls, the "crescent" is the voice of the Tasmanian mountains — the music of the wilderness.

This morning there was a snap frost and the waterfalls and creeks have all but frozen. The water barely moves, dribbling in pulses over the crystalled lips of knobbly rock. I run my hand over the stone, greeting the place that has become everything to me. Out here little changes; I find comfort in that.

Once, I came here just for the mountains; like anyone else who first ventures into this realm, I saw only the obvious. Returning again and again, I have established an intimacy with Nature formerly unknown to me. What I want to talk about here is that feeling and why it is important to me. To do this, I'll have to give away a few secrets about this special place.

Perched high in the Central Highlands is an alpine plateau, spacious and wild. Though ringed by impressive ridges and towering cliffs, the feeling here is subtle, rather than grand. It is a gentle place, a fragile place, where glassy pools are cradled by containing walls of cushion plants, and flowers sway from rock ledges. At night you are soothed to rest by the echoes of chattering frogs and the whispering trickles of a creek. This is not only the place where I began my first excursions into the wilderness, but it is also where I began to explore my own mind.

I have come to love this place as no other. It is secluded and safe, lying beyond the reach of reckless hands. This remote range, tucked away in one of the wettest corners of our wooded isle, has become almost a shrine to me. It was here that I first became conscious of the depth of wilderness, of its subversive ability to deprogram and to recast our thinking.

I generally return here alone. For me, entering wilderness accompanied means perhaps seeing only a portion of what it offers — we see the "scenery", but can miss the profundity. To be receptive to all its facets one should reflect from the perspective of solitude.

Solitude is "aloneness", as distinct from loneliness. In loneliness we discover nothing, but in solitude lies the mechanism — if we take the opportunity — for seeing beyond some of those maxims which are taken to be the cornerstones of our civilisation, yet which so often are nothing but the causes of our derailment. In a state of isolation we can evaluate, more carefully and deliberately than at other times, why we arrive at the values we have. Perhaps only in solitude can we cast off the tethers that limit our thinking.

On the slopes of this mountain lies a goblin forest of gnarled and contorted Myrtles; they are caked in moss so soft that you can't pass them by without touching it to confirm that such a thing indeed exists. The twisty forest is ideal habitat for the smaller and generally unseen birds that bob around on the scaly trunks, hanging upside down, and filling the air with the warbling song that makes these rather plain birds so appealing. On my way here, I rested among fallen branches caked in mushrooms; my exhaustion was soon forgotten as I watched a Scrub Tit flitting around on the mossy fronds, pouring out its musical trill. Once you've savoured this call, you can never turn your back on the wilderness; the bonds have begun.

I seek cover high in an open bowl — a cirque-like depression which offers shelter from the raging westerlies that go hand in hand with these mountains. Drenching fronts from this quarter dump some of the heaviest rain in Tasmania, and keep these places as they are. If you don't like the rain, you can't love these mountains.

Beneath the final summit block, trickling creeks rush from the open slopes to form a graceful tracery over the damp and spongy turf. It is a natural blending of water, rock and flowers, the essential elements of an alpine garden. I feel at my best here.

On the edge of this protected scoop, colossal shafts of unbroken rock plunge to the tinkling forests. To stand on the edge of these spooky cliffs as echoing bird calls drift up the shaded walls is one of the most calming sensations I know. The rock here is dolerite, an ancient rock from deep within the Earth. For shape, dolerite has no rival, being cast in an appealing array of columns and blocks; and in the flood of light at dawn or dusk these walls are infernos.

Whenever I return here I begin by "doing the rounds", reacquainting myself with the place. I wander round the familiar rock pools and animated boulders, letting myself open up again and confessing that I've been away too long.

In my ramblings I'm suddenly almost swept off my feet in shock — SWOOSH... I scan what's left of the sky and glimpse the blurred traces of what look like shooting stars disappearing into the gathering mist. The silhouettes — and speed — are unmistakable. If the magnificence of flight is embodied in a single bird, it is in these sleek little strangers from the northern hemisphere. Needletails (Spine-tailed Swifts) are perhaps the most aerial of birds, spending almost all their time on the wing; it is thought they may actually *sleep* on the wing, catching rest as they fall, with wings folded, before regaining their graceful flight. They are marvels of the bird world — among my favourites.

The mist has crept right down now, obscuring the other side of the bowl. When the weather sets in like this, the place feels wholly detached from everything. I'm wonderfully alone, just me and the rocks. Immersed in this comforting shroud, I imagine for a while there is total harmony in the world: how can there be so many divisions out there, when here there are none? While this feeling is illusory, the impression is real enough; it's sufficient reason to keep returning.

It's turned bitingly cold and the swifts have fled; only drifting snowflakes remain, braving the air like daymoths.

Summer or winter, the snow is always a welcome attraction. I like the calming effect of the feathery crystals coasting to earth. When it snows out here I'm excited, continually unzipping the tent door — every five minutes — just to see how things have changed.

I've always been fascinated by snow; it is, after all, a minor miracle. Many childhood memories revolve around jumping into deep drifts — head first — and coming home saturated and freezing. I used to watch it decorate everything it touched; and secretly, I still get a thrill out of the way these soft, spidery crystals of water can make almost everything unmanageable. Men shelter, physical work grinds to a halt. Snow humbles.

After several days of whiteout I venture out into the world of magic. Cornices festoon the cliff edges, framing a vista of Earth's most beautiful work, floating on fog. Swirling whorls of wind-blown ice crystals — spindrift — smoke over the soaring columns of silvered dolerite, almost obscuring the snow clouds piling above.

It strikes me very clearly now that these clouds are essentially the same clouds as those which roll over the Arctic slopes and saturate the Amazon Basin. There is a universality about them which links the wild places to us all and, for me, reaffirms my intuitive belief that dotted lines are for guarded minds; like birds wild and free, clouds know no boundaries and make a mockery of ours. Clouds also strengthen our grasp of the notion that wild places should exist for their own sake, not ours.

Suddenly a Wedge-tailed Eagle appears, drifting like a leaf over the powdered slopes; it banks and rolls among the frigid clouds, taking in everything. The sight is gorgeously wild. Swirling mist plays on the ridgetops as the bird grapples with the gusts. We eye each other, man and eagle. I try to gauge its thoughts as it looks down with those keen eyes, tearstained in the freezing wind. Don't let anyone tell you birds don't enjoy flying — that bird was in its element, and as ecstatic as I was. This magnificent raptor is a symbol of these mountains, epitomising the spirit of the wild.

When I reach the edge of the cliffs — the overlook — fog is pouring up the cirque walls like smoke up a chimney. The scene is so overpowering I can't help but let go. I race around out of control, not knowing where to begin; at a time like this, where *do* you begin? Then, before I can take any more in, beams of sunlight ignite the towers of rock, fluorescing the mist into a colour there's no name for. I laugh at my sense of abandon, but sufficient self-consciousness returns to make me look round, embarrassed, making doubly sure there's no one else here: how could I explain my childish delight? I feel, perhaps for the first time in my life, what it means to be free: the dogmas and the rules are powerless here — nothing can touch me. I am beyond the reach. I know now what the eagle feels as it hangs on fingered pinions above the funnel of light surging up the fluted rods of stone.

As suddenly as it had begun, it's all over; the mist evaporates and the magic fades to daylight. Another day begins. Apart from their biological necessity, we need these places for other reasons: to be our own selves and nothing more; to "reconstruct our individuality"; to feel alive — simply to *feel*. We need these places to illustrate the difference between existing and *living*. We need the opportunity to "let go completely" for only in this frame of mind, perhaps, can we unshackle ourselves from the conveyor-belt of conditioning.

It will take persistence and love to ensure that there will always be places remote and wild, living and breathing, places belonging beyond the reach — not only for what we can learn from them about Nature, but also for what we can learn about ourselves. They allow us to free our consciousness from the boundaries within which convention has it confined.

In wilderness — Nature's original face — lie many of our answers. By understanding ourselves we can better confront the future dilemmas which we will inevitably have to face. There is no finer place for self-discovery than in the wild corners of our Earth.

It's raining heavily as I begin my trek down to the lowlands. I leave the garden and the ponds to the hiding frogs; they're delighting in the deluge.

Before I confront the bouldered slopes I take a parting glance, letting the wildness leave its impression so that I can recall it later when I will need it most. Dragging myself away, I begin to descend.

Somewhere ahead, deep in the dripping forest, I hear the gurgling trill of a thornbill, a trusting, cheery little songster of these ownerless mountains — just one more reason I keep returning.

THE PORTFOLIOS

HIDDEN GARDENS

TASMANIAN WARATAH AND ORITES, MT THETIS

LICHEN GARDEN, MT INGLIS

PANDANI FRONDS, CRADLE MOUNTAIN

SCOPARIA GARDEN AND APPROACHING SHOWERS

FROSTED PANDANIS AND PENCIL PINES, DU CANE RANGE

WOMBAT SKULL

DECIDUOUS BEECH, AUTUMN

DECIDUOUS BEECH AND PENCIL PINES, AUTUMN

THE FOREST

PENCIL PINES AND DECIDUOUS BEECH, LATE AUTUMN

SOUTHERN BEECH (MYRTLES), ECHO POINT

AUTUMN FUNGI, MT PILLINGER

CELERY-TOP PINE LEAVES

AUTUMN FUNGI, CRADLE MOUNTAIN

HYDROCOTYLE SP. AND WOOD-GRAIN

PANDANIS IN SNOW

HARD WATER-FERN

TRIBUTARY OF THE FORTH RIVER IN MIST

FROST

CRADLE MOUNTAIN, ERRATICS AND FROSTED BUTTON-GRASS, DAWN

FROST DETAIL, WINDERMERE PLAINS

ICE-COVERED BRANCHES, THE LABYRINTH

FROSTED DOLERITE, DU CANE RANGE

MT OSSA AND FROSTED BOULDERS, DAWN

FROST DETAIL, LAKE WINDERMERE

FROSTED PAVEMENTS, DU CANE RANGE

THE HIGHEST REACHES

SEEDING HEADS OF MILLIGANIA LILY, DU CANE RANGE

MT GERYON IN SNOWSTORM

MELTING SNOWBLOCK, MT THETIS

MT HYPERION AND MT OSSA, SUNSET

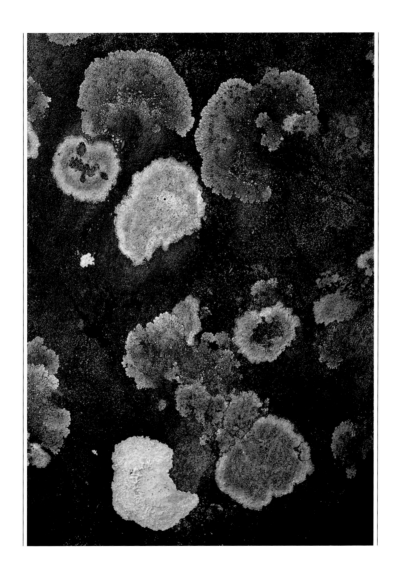

LICHEN DETAIL, THE LABYRINTH

GLACIAL BOULDERS, LAKE OPHION

EXFOLIATING DOLERITE AND MT GOULD, LATE AFTERNOON

ALPINE GRASSES AND TARN, DU CANE RANGE

MORNING LIGHT ON FROSTED DECIDUOUS BEECH AND PENCIL PINES

ROCK

LICHENED DOLERITE

RIDGE CREST, DU CANE RANGE

PADDYS NUT, EARLY MORNING

DOLERITE BOULDERS, DU CANE RANGE

LICHEN DETAIL

BOULDER GARDEN, MT THETIS

EPILOGUE: A WESTERN TASMANIA NATIONAL PARK

TYNDALL RANGE, SUNRISE

Epilogue: A Western Tasmania National Park

From the isolated peak of Barn Bluff, the country falls away steeply to the western rivers: the Bluff, the Fury and the Mackintosh. These are alluring names known mostly to people who study maps, for very few have actually ventured here. Those who have speak of a wild region with a pervasive sense of remoteness, for this country of dissecting gorges and pure stands of native conifers is one of the most remote landscapes remaining in Tasmania. Its remoteness, and the fact that so little is known of it, make it one of Australia's most precious wilderness areas.

Yet, significant as it is, it did not receive protection under the terms of the — nonetheless extraordinary — agreement reached by the Labor–Green political alliance in 1989. Substantial areas of the State were added both to the World Heritage List and to the national park system (1990); the total World Heritage-protected area now extends to 1 387 000 hectares. But, predictably, contentious areas were excluded. The area west of Cradle Mountain–Lake St Clair National Park, the Reynolds Falls area to the north, the West Coast Range and other key areas were not included for the usual reasons. A glance at a land tenure map or a geological map explains: part of one area lies within a timber concession; other parts lie within the Mt Read Volcanic Belt — a mineralised section of the west coast; these areas are designated "Conservation Area". While the title is impressive, its status is not; Conservation Area legislation still allows forestry activities, mining and dam-building.

Land tenure maps, and the regulations they embody, reveal a society's priorities. In Tasmania it is obvious where ours lie. The distant wildlands to the west, the Reynolds Falls area, West Coast Range, and the land south of Macquarie Harbour have not been included in the park system because too few of us understand why wilderness should be preserved in the first place. If, and only if, the land is "useless" will the authorities concede it to the national park system. This mentality has a name: it is called the "residuals" approach, and it is not confined to Tasmania. World-wide, similar attitudes prevail. We do not sufficiently understand why a landscape without intrusions is so valuable.

The best gift Tasmania can offer the world is its distinctive wilderness. Almost the whole surface of the Earth is roaded, ploughed, built on or lived on; the surviving open spaces are priceless. Very few "unknown" areas exist any more on our planet, yet in Tasmania there are still places where humans have *never* been.

In 1975 Unesco established the World Heritage List. Many countries, including Australia, are signatories to the founding convention. The list registers international recognition of a site's natural or cultural value. In 1982 the World Heritage Committee recognised the outstanding value of Cradle Mountain–Lake St Clair National Park and, accordingly, added it to the register. It joins such places as Mt Everest, Grand Canyon and the Pyramids of Giza.

Australia's federal World Heritage legislation is amongst the strongest anywhere. The fact that Australia has used this legislation to protect natural environments, and that nominations have often produced volatile controversies, indicates the beginning of a welcome change in our attitudes towards the natural world.

While the 1989 decision was outstanding, key areas are still unprotected and are being progressively damaged. Many people have lobbied hard for the protection of western Tasmania. The Western Tasmania National Park proposal was first put forward over 15 years ago. Though this book is about Cradle Mountain–Lake St Clair National Park, clearly the park is not a self-sufficient entity. Its boundaries are imaginary lines conceived for management purposes. If we are concerned about the future of our wilderness it is important that we see this park as a stepping-stone towards a much larger and more significant park, with rational and ecologically defensible boundaries.

The enlarged Western Tasmania National Park which is now being campaigned for boasts many impressive characteristics, and these include: the tallest flowering

REYNOLDS FALLS

CONGLOMERATE DETAIL AND GLACIAL VALLEY, TYNDALL RANGE

trees in the world, some of over 90 metres; some of the world's oldest trees, including Huon Pine which have been dated at 2 600 years; the longest and deepest cave systems in Australia, many with unique species of invertebrates; the most extensive collection of freshwater lakes in the country; among the oldest archaeological sites, and the most outstanding glacial landscapes in Australia.

But as the years pass the opportunity to preserve this extraordinary landscape will slip away. The national park movement has been one of the most important phenomena of the twentieth century, and has been a catalyst for people's increasing attachment to the land. But this attachment needs to be both deeper and more widespread — the land needs to be seen as something more than simply a "resource". When we accept it as an important part of our being, it ceases to be merely a surface that we look at, drive across, or walk into, but rather it becomes a profound link with the Earth, deeply rooted in our psyche.

The monumental forces of snow and ice moulded and carved the original Tasmanian landscape, perhaps beyond recognition. Humans, however, are changing it even more dramatically — and faster. The amount of love we bestow on our land will determine whether our species will endure or whether we will be as transitory as the ice.

ICE-ABRADED CONGLOMERATE PAVEMENTS, TYNDALL RANGE